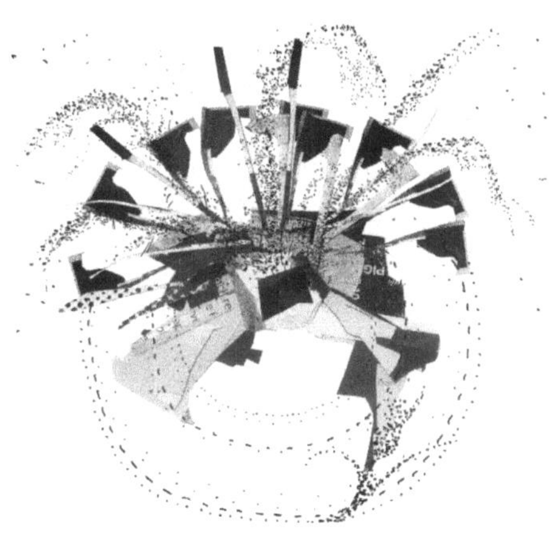

Ontologies of Environmental Collapse: Internet Energy Intensity and Efficiencies
Volume 2

© Nuala Loges, 2021

𝕴𝖓𝖘𝖎𝖉𝖊 𝖙𝖍𝖊 𝕮𝖆𝖘𝖙𝖑𝖊, ITC-030

Typography design for this volume was inspired by the elegant details of Continent journal. Thus the text is set in Frutiger LT Std, a variant of the Frutiger typeface often used on pharmaceutical labels due to its legibility. Adrian Frutiger's life was filled with tragedy, losing his first wife in childbirth and the two daughters of his second marriage to Selbstmord. His sans-serif typeface is used in Amtrak branding. Titles are set in Memphis Light.

Cover design was inspired by the xerography of Deepskin Conceptual Mindmusic cassette releases of Clinton Williams' experimental electronics act Omit.

Both of these touchstones were curated by Maure Coise to enhance the physical characterization of these volumes and their congruency with the mission of the text.

Designed by John Trefry for **𝕴𝖓𝖘𝖎𝖉𝖊 𝖙𝖍𝖊 𝕮𝖆𝖘𝖙𝖑𝖊**

ISBN-13: NO!

Ontologies of Environmental Collapse is a text in the expanded field of literature. Visit www.insidethecastle.org for more.
𝕴𝖓𝖘𝖎𝖉𝖊 𝖙𝖍𝖊 𝕮𝖆𝖘𝖙𝖑𝖊 is located in Lawrence, Kansas

Ontologies of Environmental Collapse: Internet Energy Intensity and Efficiencies Volume 2

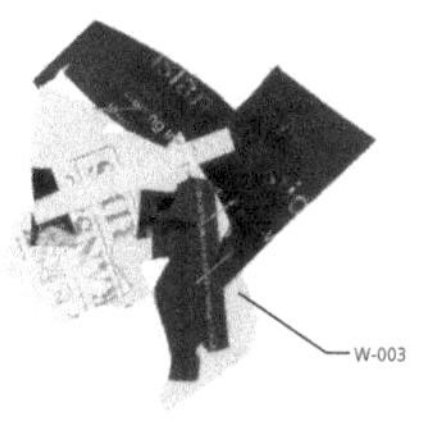

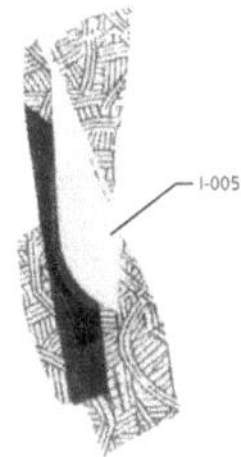

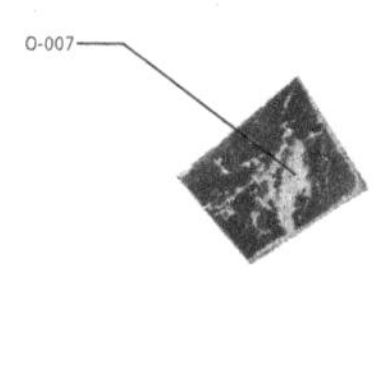

Nuala Loges

also in **Internet Energy Intensity and Efficiencies**
Volume 1: Cultural Object Ontologies by Maure Coise, 2019
Volume 2: Ontologies of Environmental Collapse by Nuala Loges, 2021
Volume 3: Structuring Information by Maure Coise, 2023

Spring 2021, Lawrence, Kansas, 110pp., $13

Ontologies of Environmental Collapse

Standards are No One's

On a young river, shale breaks. Deposited
within sight of swimming figures, diving.
Passing by pleasure boaters. Submerging
in the water, the rocks become invisible.
There's a rhythm in the swimmer's bobbing.
Far off there are ripples and some birds cry.
The Scioto River, it's tributary,
the Oletangy River, and their banks
have always been sites of pleasure. Around
two million years ago they were formed,
along with the Ohio River, by
glacial melt. They, along with the city
of Columbus, are in a glaciated
plateau. Today, corporate complexes
are dotted around the ancient Great Serpent
mound.

Everyday the land is further domesticated.
Cut against the mounds built two thousand years
ago are concrete and glass offices
and bird cries.

I am a library worker. I no
longer live in Ohio, but I was
born there. I am a settler on the continent
and am concerned about the physical
space my work takes up.

The Ohio College Library Center,
later changing their name to the Online
Computer Library Center, formed in
Columbus in 1967.
They manage WorldCat, the largest bibliographic
database for shared cataloging and
public access. Cataloging is here
defined as metadata design, human
and machine-readable aboutness, practiced
by defining attributes and hierarchizing
relationships for description. 'Aboutness'
means the shorthand coding of the subject
of material in categorization.
Once a record or bound description has
been made, it can be made available
for others to integrate into their
own systems via WorldCat.

I'm looking for a way of writing environmental
justification to guide bibliographic
ecologies. Literary description,
allowing for multiple senses of
a word like ecology, to describe
the relationship of cataloging
standards and environment, and expropriation,
outside the conventional environmentalist
purview. The topic of 'wildness' is exemplary,
making the links between "posthuman" ethics
and accounts of agency within Western
imperialism. Anti-colonialist
theorizing rests on prior social
research in "postcolonial" critique.
The importance of imagination
for ontologies and Ontology
will be elaborated through this poem
in the context of decolonizing
methodologies. My approach is inspired
by the relation of literature
with sciences, and design, elaborating
mutual care and new sensibilities
for the Information Sciences. To
provide a better foundation for a
cataloger's recognition of their
responsibilities in colonization
and exploitation, I weave considerations
of OCLC and WorldCat with Indigenous
politics.

To begin, imagine the data centers
storing WorldCat hum with the sounds of Computer
Processing Units. Imagine the colors.
What are the materials of construction?
How does the sky look outside? What is it
like to walk along the halls? How is it
similar to swimming in the river?

I do not claim to be providing a
treatment of colonialism here,
just as I do not claim to be doing
actual decolonization work.
I mean to advocate for anti-colonialist
informatics practices. I am offering
a critical view of environmentalist
approaches to data and information.
Regulating electricity and
water use is not enough.

Lingering justifications for colonization
are implicit in many current discussions
of environmental sustainability.
The invisibility of these justifications
binds environmental consciousness to
particular conceptions of time, so
that homogenizing, colonizing
regimes can continue, even with the
best intentions.

Corporate interests obscure the real relation
of computing and environment. Data
center companies make use of nature,
caught up in actual space with long distance
impacts that can be traced through mass killing.
For there to be any reconciling
with those being marginalized and killed,
I propose that those attempting control
of information effectively cede
power. Stepping down from control of information
infrastructure seems necessary because
sustainability is impossible
under Capital.

For example, the state of Ohio
privatizes information as an
aspect of its culture. By exploiting
agents and fitting data into an
information regime, Ohio
is essentially extending corporate
practices and furthering the project
of expropriation. The physical
infrastructure of data centers situated
in particular places, implicating
any number of agents that comprise
place.

Using WorldCat, access paths pass through places.
Information access can be situated
in displacing. OCLC, as an
extension of the colonial project,
is not expelling human bodies, but
controlling the means of access to
information. WorldCat's legitimacy
is inextricably bound to larger
information regimes. The work is performed
with barriers and discrimination.
It is possible to emphasize resistance
by information workers within this
framework, finding ground by acknowledging
the physical labor involved at OCLC
and its data centers.

After the End of WorldCat

A library cataloger describes
an entity for a retrieval system
by encoding information in MARC
fields. Descriptions are input through software
developed by OCLC, called Connexion,
to upload directly to WorldCat and
to the specific library's database
by using software, such as Ex Libris's
Alma. Variations in formal descriptions,
even for such seemingly basic information
like title or author, are actually quite
complex, and are accounted for by numerable
MARC fields, that also vary in degree
of formalization.

Records written using this accidental
ontology are available to
other librarians for copy cataloging,
saving time and enhancing discoverability
of previously undescribed resources.
I say accidental, because what was
intended only to automate card
cataloging has become a standard
for formatting data, an environment
for knowledge representation. But, this
data environment is not really
comprised of 'standards', formalizations
of 'agreed upon' rules for describing.
Although sometimes ignored in acts of describing,
data structures are physically made
up of electricity and computer
hardware.

How Are Information Ontologies Possible?

Although OCLC's WorldCat appeals
to an inclusive posture, public access
to records is built through expropriation,
and justified by corporate tactics,
privileged logics. Grounded in exploitation
of particular contracts, information
infrastructures exploit actual living
conditions. For catalogers, agents
are considered resources, engaging
with a colonial, extractive attitude.

OCLC stores their data along
the Scioto and Oletangy rivers
in data centers in suburbs of Columbus:
Dublin, Ohio. European settlers
claimed this land and forcibly removed Indigenous
persons. The Shawnee, along with the Ottawa,
Potawatomi, Sauk, Meskwai, Kickapoo,
Ojibwe, Odawa, and Miami
tribes, were relocated to Oklahoma.
Although the Hopewell mounds have achieved status
as a National Historical Park,
less memory is paid to the Great Lakes
tribes.

Despite wars fought for multi-state solutions,
the land was subsumed by the United
States. Throughout the 19th century, many
ways of living and of knowing were systematically
eradicated. These actions continue,
avoiding accountability for
systematic oppression of ways of
living and of knowing.

Columbus dammed the Scioto in 1967.
Since 1987 water turbines
draw electricity for the region.
Columbus owns the dams. Flows of money
accumulated in the city of
Columbus. This city has its own division
of power. The turbines closed for renovation
from 2015 to 2022,
during which time the municipal service
returned to powering by fossil fuels.
Hydroelectric power, often presented
as an environmentally friendly,
renewable power, evidences
a massive conflict in the concept of
the environment. Extractivism
is embedded in the very concept
of an environment as a natural
resource. Ignoring this, all environmentally
friendly motivations are suspect. What
the Anthropocene calls for is abolishing
the notion of the Earth as non-agential
and exploitable.

Consider not only the point of view
of any particular information
worker, but the perspective of Information
Science. How does Information Science
see and articulate appropriate
land use? How does it see and articulate
appropriate use of information technology?

It is viewed as something to be dominated,
a concept inherent in the history
of anthropogenic agriculture
and subjectivity. However, I
am not interested in arguing
for the abandoning of agricultural
and forest management practices, nearly
ubiquitous across human cultures,
but for critiquing their anthropocentrism.

Management of power by the city
government is based on assumed land ownership.
People and environments were displaced
because of covetousness. An environmental
history focusing on this question of
ownership is useful in beginning
to describe one of many stories of
hydroelectricity that is also
a story of privatization, a
term abstracted from actual emotions
that motivate it.

By considering these emotions, I'd
like to allow a broader conversation
to take place in the Information Sciences.
Covetousness cuts deep into the concepts
of natural resources and ownership,
excessive taking. Complicity with
the state/city energy policies
means complicity with extractive regimes.

Although treaties were made, like the Jay Treaty
of 1794, the Ohio
River Valley was effectively stolen.
The least that can be done concerning an
ethical action is formulaic
acknowledgement of Shawnee tribes and families,
though this remains a potential aid to
colonizing corporate interest. Formulaic
acknowledgements can sometimes be a way
of eliding action and is not necessarily
the best way of showing up for real problems.
My interest is in advocating
for multiple local ways, or Indigenous
information systems.

I want to draw a correlation between
collecting cataloging records and
underlying exploitation of natures,
so as to consider cataloging's
futures. My advocating here is not
meant to make claims from the outside about
Indigenous politics, or assume
that in some way Indigenous sciences
will "redeem the whites". I am gesturing
towards some change in conceiving the relation
of information and environment,
physicality and categorization.

Categories' Gathering Place

OCLC's shared cataloging data
sits on land that was acknowledged as having
agency by its Indigenous inhabitants.
This transformation was realized through
violence, expropriation, genocide
and slavery. Though there are many ways
to unpack definitions of violence
in a knowledge organization context,
I will focus on colonial oppressions within the
technological structures of cataloging,
in terms of their physical embodiment.
Even climate modeling and meteorological
simulations have physical impacts.
Data use can shift the site of pollution.
A computer lab can become 'greener'
by switching from PCs to running virtual
machines and saving data on the 'cloud',
however, this only means that the data
is stored elsewhere.

Pollution in the soil, air, and waters
of Ohio has accumulated
since the 1800s. Many larger
coal-burning power plants are, since the 2010s,
closing or downsizing. Many have consistently
failed to meet the Environmental Protection
Agency's standards, just as many of
the cities in the state fail to meet air
quality standards.

Every time one uploads an image,
streams a video, or even performs
a web search, electricity is used
to store the data and transmit information.
I raise this issue as imbricated
in a critique of library sciences
as a violent sense of ownership
in relation to how we know ourselves
and relate to the Earth. This might serve as
a general description of the Internet,
beyond library cataloging systems.
Taking up space, just as any platform
for instant gratification and self-knowledge
or self-expression entraps one in patterns
of oppression. The platform users become
implicated in the platform's taking
up space.

The existence of data in server
centers uses electricity, and
also a substantial amount of water
for cooling.

Providing information sciences
with a concept of applied ecological
knowledge might use a cultural value-
based approach.

Excessive use of "resources" benefits
some at others' expense. The idea of
extractivist regimes, particularly
through histories of abuse of Indigenous
populations, offers an idea of a care assemblage
to guide activists. 'Care' I use in the
broadest possible sense. A philosophical
idea of orders of epistemological
exclusivity, from the organization
scheme to its historical development
as a system, the world of shared knowledge
systems maintained by cataloging technologies
is compromised due to inadequate
system effects, that I interpret as
epistemic oppression. The exclusionary
spaces of corporate complexes, privileging
"good business/living", historically, expropriate.
Domestication and political
control are inextricably intertwined
in the history of expropriation
and its present-day forms. It seems useful,
in further considering the design
of OCLC's information systems
to develop and design pluralistic
ontologies for multi-agential
use, decolonizing relations of
entities and acknowledging, or loving,
the agencies of such entities and
natural forces will play an important
role in developing these ontologies.

Reckoning with multiple ways of knowing
in actually existing communities
is problematic in Ohio. I
do not know if this reckoning is a
question of land ownership or ethnographic
identity. I consider Information
Science as it is subsumed within the
culture of colonization, a culture
that inherently includes a violent
sense of ownership and excludes possibilities
of multiplying identification.

To consider the possibilities
of improving cataloging technologies,
to structure information and information
access in a way that is less consumptive
and less colonizing, I recommend
studying Indigenous information
retrieval systems. However, my point
here is more indirect, recommending
information workers to think about
how they abstract their work from its physicality.
The social-environmental impacts
of any ostensibly purely conceptual
description can be inferred from how information
professionals think about the physicality
of their intellectual labor. This
might be considered, in a general
sense, as abstracting concepts from environments.
Information Science and Technology
creates barriers when perspectives are
excluded, and healing ought to attend
more to social contexts.

"Greening" Information Organization

Corporate practices making up sustainable
agendas may or may not reduce pollution.
'Greening practices', concerning cataloger's
accountability, are suspect. CeraNet,
the company whose data centers OCLC
uses in Ohio recently won
the Columbus GreenSpot award and American
Electric Power Ohio's Green award.
In regards to green policies, OCLC's
data centers winning an award from
the city of Columbus or AEP
Ohio is more of a concern than
a relief.

I write this to unsettle the legitimating
of land use as permitted by the State,
to further explicate healing. Tracing
flows of money might provide another
useful way to instigate social-environmental
change within Information Sciences,
unfortunately all physical data
centers, and the relations of cataloging
institutions and their physicality,
are maintained with a degree of secrecy.

Still, decisions needed from information
workers become clear. What is really striking
is how, within the current regimes, there
is no room for emotional relations,
or, there are of course emotions, but they
overwhelm, and are ignored. How can catalogers
respond to the often horrific excess
of Capital itself? Reckoning with
the mass death of plants and animals as
a direct result of certain activities
cloaked in the good intentions of information
science and sharing knowledge, surely stirs
strong emotions with those catalogers
who recognize the impact. And then how
can these workers contend with important
variations in degree of vulnerability
correlating with cultural variations?
I encourage library workers and
users to work towards opening space for
the legitimacy of decolonizing
methodologies, by analyzing ever
more the social elaboration of
information work. An ecological
approach to information science is
the next step to an environmentalistic
concept of information.

Bibliographic Ecologies and Ecological Knowledge

A long time ago, on the east coast of
North America, where the Potomac
flows down from the Chesapeake Bay, Uttapoingassenem
came down and centralized the governance.
Deep in the waterways, where hardwood forests
met wetlands, Uttapoingassenem
organized a system of chiefdoms through
focusing on the organizing of
forestry techniques around 1658.
Not long after that the Piscataway people
of the North fully replaced the Anacostans.
This region is subtropical in summers,
with hardwood forests composed of ironwood,
ash, sassafras, and pawpaw. One can also
find spicebush, summersweet, hazel alder,
and silky dogwood. English colonizers
used this land for their own capitol, Washington, D.C.,
an effort to centralize a different set of practices
and people. The city was designed with
little attention paid to the surrounding
environment or forestry techniques
that had been vital to the now colonized
people's organization. It stands out
as an outpost of pseudo-Roman architecture,
facing the Atlantic.

Since the beginnings of the colonial
empire, Indigenous people and
their ontologies have been systematically
and haphazardly crushed. Though using the
phrase, "Indigenous ontologies"
is generalizing, referring to
pan-Indigenous movements is in no
way derivative of European
universalism. In fact, it is
a political frame in which to critique
imperialist universalism.

Considering Indigenous ontologies, is BIBFRAME
moving toward a structure of multiepistemologies?
The organizing of diverse cultures
into an autonomous pan-Indigenous
or pan-Indian movement is a response
to de-legitimization of tribal
self-determination. Political
fragmentation is one model for 'multiple
epistemologies,' in that the fractures
(might) represent substantive bifurcations
from a previously homogenous
movement. Place-based actions of sovereignty,
having to do with self-determination,
land use and particularity of cultural
practices, contain the potential for
loosening the negative connotations
associated with divisions of
geopolitical identity.
As geopolitical movements consolidate
in order to mobilize, appeals for
a nationalist unity help the
State structure 'the people' toward realizing
State interests and goals. I'd like to consider
how to counteract the homogenizing
effects of nationalist growth.

False unifying concepts, such as the
very notion of categorizing
persons with a term like 'Native American',
simplify identification and
ways of living and knowing. The placement
of 'Indians of North America'
in the Library of Congress Subject
Headings under 'History', the argument
offered by the Library of Congress
was that this was necessitated by
use, that the expertise on the subject
was somehow not the living 'Indians
of North America'. It is not just
the Library of Congress upholding
colonialist racism, but its critics
who raised the issue within non-Indigenous
scholarship who assumed that Indigenous
ontologies could possibly be assimilated
into a standard ontology at
all upholds the same racism. Whether
at the Federal or State level, the
dominance asserted by categorizing
systems subsumes lives into resources
for the systems' purposes.

Philosophical accounts underpinning
distributive justice can be shown to
underpin Information Science as
idealization. This calls for a critique
from within Information Science, and
a critique in a non-ideal sense, considering
many aspects of institutional
practices constructing information
retrieval without assuming an underlying
rationality to the agencies
or social phenomena.

Our job as information professionals is to recognize
the various forms knowledge organization
work can take, and find ways to support these projects
out of respect for the decolonizing
and self-determination efforts of
Indigenous communities.

The economy is a driving force in
developments within Information
Science; when financial interests pilot
any project, certain people and other
entities are always marginalized.
In Information Science, what are assumed
to be formal procedures do have physical
instantiations. Through questioning natures
and systems of knowledge, in response to
assessments of physical data centers
of library systems, we may approach
openings in the rhetoric that facilitate
changes within these systems and consider
the semantic of business policies
and environmentalism. Since Spanish
colonizers entered and occupied
Anahuac, setting up the modern state
of Mexico, text-based systems of knowledge
were privileged, excluding
the information of ceremonies,
songs, dances, stories, hunting and growing
and healing practices, painting, pottery,
weaving, carving and imaginal work.

The Physicality of Virtualization

Appraising corporate methodology
might, on one hand, provide a platform with
which to question primary justification
mechanisms within political
instantiations. But also, and importantly,
these methodologies utilize 'green-washing'
language to provide
ethical justifications and skirt some of the
very issues I seek to unveil. Green-washing
language within corporations is responsible
for the blurring the edges of decision
making, creating a façade of ethicality
where financial motivations are of
primary concern. This normalizes
exploitative practices, stifling debates
over epistemic problems embedded
in colonization. OCLC,
or the Library of Congress in Washington D.C.,
or differently, and relatedly the
Israeli company Ex Libris, are
developing on territories through
oppressive acts.

Is it possible for Information
Science to look within itself and make
explicit how expropriation and
forced relocation takes place, and how this is
a tool for regulating a wide variety
of cultures and localized ways of living
in reciprocity with environments?
Particular ways of knowing have been
dismantled through the forced reliance on
State property rights. The environmental
impact of particular Information
Technologies situated in the extractive
regime of American expansionism,
a story of exploiting land for agriculture
and mining, through which any discussion
of ecological sensitivity
can be filtered, can be treated in any
intersections of sovereignty, cultural
practices, and computing/informatics.
 "Justice" here can be just as easily
analyzed in pre-digital techniques
of knowledge organization, digitalization
merely continues in marginalizing
ethical difference through particular
business policies and investment in power, through infrastructure and
epistemologies.

Reliance on extractivism is a
survivance issue. To assess, to ask
where the responsibility lies, and
who profits, what local economies
have been decimated; one can look to
corporate policies. There are rarely
any internal policies on environmental
standards for cataloging standards. For
example, a Stanford University
employee ostensibly follows: Stanford
University's Energy and Climate
Plan (2015), Santa Clara County's
Sustainability Master Plan (2017),
California's Green California State
Administration Manuel (2016),
the United States Environmental
Protection Agency's Clean Air Act (1963),
Clean Water Act (1972),
Resource Conservation and Recovery
Act (1976), the United
States Congress's Green New Deal (202x),
the United Nations Framework Convention
on Climate Change Kyoto Protocol
(1997), and the UNFCC
Paris Agreement (2015).
Likewise, the Library of Congress employee
follows: US Office of Personnel
Management Fraud, Waste, Abuse policy
(2016), the Sustainable
DC Act (2013) as well
as the national and international
policies.

Survivance, as continuation
of stories not individuals, seems
to be one adequate frame to account
for ontological pluralism
in its full philosophical and religious
senses, and as inherent in spatial
politics.

Many institutions are actively
working towards future systems, collecting
data, and making environmental
impacts. Unfortunately, in the case
of experimenting with transitioning
out of MARC and into BIBFRAME, everything
is cataloged in both MARC and BIBFRAME,
effectively doubling environmental
impact.

It is also important to emphasize
that the reach of the BIBFRAME experiment
is in fact limited. The development
of BIBFRAME occurred as an initial
pilot between 2014-2017
was organized with: Biblioteca
Nacional de Cuba "José Martí,"
Columbia University Library,
Cornell University Library,
Library of Alexandria, Music
Library Association, National
Library of Medicine (USA),
Princeton University Library,
University College London Department
of Information Studies. A second
pilot from 2017-2019
included: @CULT, Colorado College
German National Library, Ex Libris/Alma,
Library of the Hungarian National
Museum, ReasonableGraph, Stanford University,
United States Army Corps of Engineers
Research and Development Center Library,
University of Illinois at
Urbana-Champaign Library.

Initially catalogers in the
BIBFRAME pilot described resources in MARC,
and afterwards in BIBFRAME. They flipped this
sequence in between; and are now planning
to have the catalogers describe resources
in BIBFRAME only, and convert the data
to MARC. The Library of Congress must
continue cataloging in MARC to
support national libraries, and experiment with future
ontologies. If BIBFRAME is used in
distribution across larger repositories,
there need be less duplication in local
caches of library institutions. Going
back and forth between them evidences
efficacies and efficiencies
of each data format.

An automatic converter using
XML Stylesheets does not solve the fundamental
problem. eXtensible Stylesheet Language
Transformation files are used to transform
an XML schema/instantiation
into, for example an HTML
page. This is the program that, in experiments
to automatically convert a
BIBFRAME record into a MARC record
and vice versa, allowed, differently
from in the initial pilot program,
the cataloger feels like they are only
cataloging one record or bounded
description. However, even without the
feeling of waste, two files continue to
exist in a data center.

The inevitable costs of developing
metadata standards through duplicating
data are symptomatic of larger
issues. That the data servers sit on
colonized land.

The Library of Congress's own data
centers are in Capitol Hill. They are
called the Primary Computing Facility.
This is powered, by power that comes
from somewhere, what is officially called
the Uninterruptable Power Supply.
In 2017, the Library
of Congress started working with Accenture,
a Dublin, Ireland, based management
firm to develop new data centers
in Northern Virginia. This distance from
Capitol Hill is ostensibly to
avoid the frequent shutdowns for security
checks, as well as to overhaul the hardware
and software of the system. The government
negotiated a 27 million dollar contract
with Accenture. The companies that will
facilitate the network and monitor
the services at the new data center
are still under negotiation.

The exact location of the data
centers is also still under discussion.
What rivers will be used, their flows, and their
potential contamination will be
determined in these negotiations.
The physical displacement of Indigenous
persons from the D.C. area was
driven by a concept of the environment
as resource, for the harnessing of power.
To be explicit, my first solution
is not conceptual, but for library
systems to divest from dependence on
fossil fuels. However, this is not to
accept primary alternatives to
burning coal to power the web, hydroelectric
and nuclear power stations.

Data centers are such significant
energy consumers, that they are reconfiguring
supply chains. Servers produce so much heat
that they are used in some European cities
to heat homes. Not being deluded here
by technofantasies of smart cities,
to take server centers as an object
for critique, following a materialist
account of infrastructure, the problem
of electricity and of cooling
water is not only in their excess,
but in their sourcing, 80% from
coal. Critiques are often quick to valorize
waste; making it into a commodity
for heat recycling is symptomatic
of the imbrication of production,
consumption, and knowledge.

I do not mean to valorize access
to information, or visibility
or transparency solely, especially
in the face of catastrophes. My question
is of meaningful access to knowledge.

BIBFRAME's history is a history
that is subsumed by a broader environmental
history of the United States government's
westward expansion. To grasp the work of
the Library of Congress, its relocation
of data centers to Northern Virginia,
and also OCLC, or any
number of universities, we must
look closely at the links between today's
concern with environmental auditing
and the scope of industrialization.
A history of expanding for resource
extraction carries with it a frontier
mentality, which further brings fixed ideas
of ownership, storage, and use rights.

Situated Cataloging

Within the story of contemporary
cataloging technologies and how
they became possible, militarism,
which was only exacerbated
by the extension of the empire
across the planet in the 20th
century, is highly involved. From late
1800s into the 1950s,
Indigenous culture was ostensibly
unlawful in North America. Since
the end of the "prohibition", sovereignty,
or self-determination, has been bound
and defined by European laws.

There are over 600 distinct tribes
within U.S. political borders
alone, each one representing unique
ways of knowing, languages, and
histories. In theory, if every tribal
government had a library of their own,
organized according to the local
Indigenous epistemology
or epistemologies (in the case
of multiple peoples in one region),
we would have over 600 distinct
Indigenous knowledge organization
systems.

Overwhelmingly, systems of knowledge
are being interpreted in libraries,
archives and museums from monoepistemic
views. Organizing knowledge entails organizing
with language categories, decolonizing
categorizing, as an environmentalist
intervention in cataloging, is
limited.

If certain Indigenous cultural
practices symbolically represent
in a way that is not property oriented,
in what is now inappropriately
called aesthetics, and radically, if more
damage may have been done to traditions
since the end of prohibition, then it
can be generalized that complex stories
have become, unfortunately, filtered
through European concepts of public
and private, damaging not only the
expression of family or tribal
belonging in representation, but
also the involvement of the land itself.

This is radical because it goes against
the status quo understanding that taking
away voice is the most damaging oppression.
Eurocentric legal structures are otherwise
from the plurality of Indigenous
politics, even if this plurality
is mobilized in a unified front.

A thesaurus for user-centered research
might do well to localize terminology.

By remembering the importance of
words in communities, we can begin
to construct naming systems that better
reflect Indigenous worldviews and assist
with data curation for those communities.

With the heritage of pollution and
devastation of the environment,
are we to continue with pride in the
legacy of Information Science
and Technologies? The electrification
of a unified nation is a racialization
driven by human choices in favor
of anti-humanist commercialism.
Are information professionals a
part of this electric pulse or can they
leverage their positions to feel this pulse?

The masks of the extractive industries
are comprised of their promises of employment
and sharing knowledge.

The horizons of cataloging can
be explored further to develop futures
that are more responsive to local concerns.
Responding to human waste's role in climate
change, certainly acknowledging responsibility
for waste, by recognizing ways of thinking
that are not oriented towards Industrial
development; environmentalism
requires no justification, as
there is no justifying a sacred
place. No entity could act as tribunal.

Continued marginalizing of particular
persons is implicit in corporate
plans for environmental and social
justice. Current information systems
are easily traced to colonial
violence. To confront such violence
and marginalization, thousands of
ontologies are needed, derived through
recognizing how people live, and different
ways of dividing natures. The weight of
acknowledgement of genocide can't be
recorded. The real gestures are unknown.

How can there be some return to Uttapoingassenem's
approach? Was this even appropriate?
Or, is this appropriate for non-Piscataway
persons to discuss at all? How to listen
to the natural flows?

One might consider, rather than worker
resistance, or the demolishing of
data centers holding catalogs by
coordinated information professionals
and replacing them with gardens, a revitalized
discussion of first principles. 'Letting
a thousand ontologies bloom' means introducing
diversity in imagining technologies.

The dominant concept is a tendency,
evident in ecology, as well
as modern technology production,
to treat a dualism of life and
machine. Systems theories, or predictive
models in particular, fail in certain
critical areas. The goal of breaking
down boundaries between technologies
and environments will be achieved through
considering cultural characteristics,
but this needn't necessarily involve
the conflation of living dynamic
environments with systems. Cultivating
non-standard theories of a duality
of technology and life can help explain
certain phenomena, such as Internet
addiction, an inherent problem in
contemporary information organization
(in this pathology, a so-called non-living
thing organizes de-humanization).
The question of how organizing emerges
is open, and the broadest possible
phrasing of the problems explained here. In
the case of improving web search functions, one
idea is to introduce more "irresponsible"
machine learning through genetic programming,
to introduce more unknowns. However,
what I want to propose is more of a
discussion of episteme, a critique
of totalization, a tendency
of enclosing multiple ways of
knowing and multiple ways of being
into a global time axis can be
countered by cultivating diversity,
honoring situatedness in virtualization.
I hesitate to call for alternatives
and frame it as a competition, but
cultivating diverse ontologies
and metadata schemas might be a way
towards self-determination of contexts
on the web.

Lived experience determines appropriateness
of standards. There is a need for increased
awareness of the laws and governance
limiting self-determination. The
role of marginalized persons ought to
be considered foremost in this challenge
of reconciling with continued development
on expropriated land. This is a
question of how persons can be considered
at all. In particular situations
that create the parameters of 'self-determination,'
the State and other businesses' actions
are more easily analyzed. 'Self-determination'
itself is articulated in Eurocentric
legal terms. Dialogues are constrained by
individuals. Self-determination
can be approached, in a way without its
prefix, through direct actions that are also
historical, acknowledging places
as embedded in systematic meaning-making.

Affirming the importance of lived experience,
I cannot imagine an Indigenous
ontology, but I can recognize
violence against this possibility.
However, recognition is here qualified;
not only are gestures such as verbal
or written (or institutional) acknowledgements
unsatisfactory compared to actual
reparations, but the gestures of colonization
and the extent of environmental
degradation are incomprehensible.

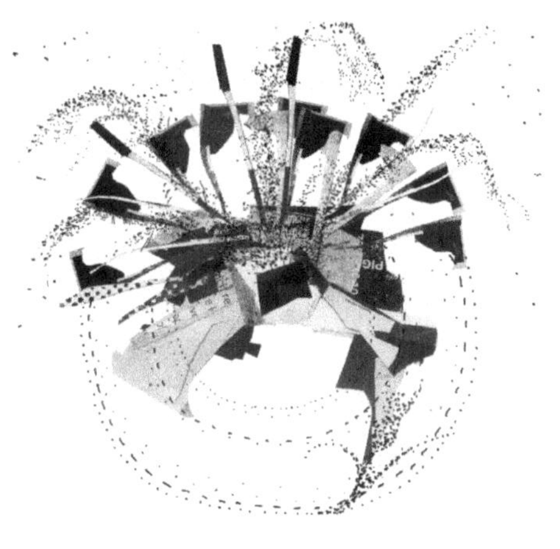